Pinaki Satpathy

Design and Implementation of carry select adder using T-Spice

Anchor Academic Publishing

Satpathy, Pinaki: **Design and Implementation of carry select adder using T-Spice,**
Hamburg, Anchor Academic Publishing 2016

Buch-ISBN: 978-3-96067-058-2
PDF-eBook-ISBN: 978-3-96067-558-7
Druck/Herstellung: Anchor Academic Publishing, Hamburg, 2016

Bibliografische Information der Deutschen Nationalbibliothek:
Die Deutsche Nationalbibliothek verzeichnet diese Publikation in der Deutschen
Nationalbibliografie; detaillierte bibliografische Daten sind im Internet über
http://dnb.d-nb.de abrufbar.

Bibliographical Information of the German National Library:
The German National Library lists this publication in the German National Bibliography.
Detailed bibliographic data can be found at: http://dnb.d-nb.de

All rights reserved. This publication may not be reproduced, stored in a retrieval system
or transmitted, in any form or by any means, electronic, mechanical, photocopying,
recording or otherwise, without the prior permission of the publishers.

Das Werk einschließlich aller seiner Teile ist urheberrechtlich geschützt. Jede Verwertung
außerhalb der Grenzen des Urheberrechtsgesetzes ist ohne Zustimmung des Verlages
unzulässig und strafbar. Dies gilt insbesondere für Vervielfältigungen, Übersetzungen,
Mikroverfilmungen und die Einspeicherung und Bearbeitung in elektronischen Systemen.

Die Wiedergabe von Gebrauchsnamen, Handelsnamen, Warenbezeichnungen usw. in
diesem Werk berechtigt auch ohne besondere Kennzeichnung nicht zu der Annahme,
dass solche Namen im Sinne der Warenzeichen- und Markenschutz-Gesetzgebung als frei
zu betrachten wären und daher von jedermann benutzt werden dürften.

Die Informationen in diesem Werk wurden mit Sorgfalt erarbeitet. Dennoch können
Fehler nicht vollständig ausgeschlossen werden und die Diplomica Verlag GmbH, die
Autoren oder Übersetzer übernehmen keine juristische Verantwortung oder irgendeine
Haftung für evtl. verbliebene fehlerhafte Angaben und deren Folgen.

Alle Rechte vorbehalten

© Anchor Academic Publishing, Imprint der Diplomica Verlag GmbH
Hermannstal 119k, 22119 Hamburg
http://www.diplomica-verlag.de, Hamburg 2016
Printed in Germany

ABSTRACT

Adders are the basic building blocks of any processor or data path application. In adder design carry generation is the critical path. To reduce the power consumption of the data path we need to reduce the area of the adder. Carry Select Adder is one of the fast adder used in may data path applications. The proposed design is implemented without using multiplexer and RCA structure with Cin=1. Instead of multiplexer and RCA Cin=1 structure hear we used simple combinational circuit. Power dissipation is one of the most important design objectives in integrated circuits, after speed. As adders are the most widely used components in such circuits, design of efficient adder is of much concern for researchers. This paper presents performance analysis of different Fast Adders. The comparison is done on the basis of three performance parameters i.e. Area, Speed and Power consumption. We present a modified carry select adder designed in different stages. Results obtained from modified carry select adders are better in area and power consumption.

Keywords: CSLA, COMBINATIONAL CIRCUIT, ADDER, BEC, D-LATCH.

TABLE OF CONTENTS

ABSTRACT 1

LIST OF FIGURES 5

LIST OF TABLES 6

LIST OF ABBREVIATIONS 7

CHAPTER 1   INTRODUCTION

    1.1   Introduction 9

    1.2   Literature survey 11

    1.3   Objectives 12

    1.4   Thesis outline 12

CHAPTER 2   DESIGN OF FULL ADDER USING DIFFERENT LOGIC

    2.1   Full Adder 14

    2.2   STA CMOS FA 15

    2.3   PTL FA 17

    2.4   TGA FA 19

CHAPTER 3   DESIGN OF MULTI- BIT FULL ADDER USING DIFFERENT LOGIC

    3.1   4-bit full Adder 21

    3.2   8-bit full Adder 24

CHAPTER 4 DESIGN OF CARRY SELECT ADDER IN 4-BIT

    4.1   Carry select Adder 27

    4.2   Conditional sum Adder 28

    4.3   Circuit description 29

CHAPTER 5 SIMULATION RESULTS

    5.1   Simulation results for FA 31

    5.2   Simulation results for multi bit full Adder 34

CHAPTER 6 FUTURE WORK AND CONCLUSIONS

    6.1   Future work 36

    6.2   Conclusions 36

REFERENCES 37

LIST OF FIGURES

FIGURE NO.	NAME OF THE FIGURE	PAGE NO.
2.2.1	Circuit diagram of CMOS	15
2.2.2	Schematic diagram of STA CMOS FA	16
2.3.1	Circuit diagram of PTL FA	17
2.3.2	Schematic diagram of STA PTL FA	18
2.4.1	Circuit diagram of TGA FA	19
2.4.2	Schematic diagram of STA TGA FA	20
3.1.1	Circuit diagram of 4-bit FULL ADDER	22
3.1.2	Schematic diagram of 4-bit FULL ADDER	23
3.2.1	Circuit diagram of 8-bit FULL ADDER	25
3.2.2	Schematic diagram of 8-bit FULL ADDER	26
4.1.1	Circuit diagram of 4-bit CARRY SELECT ADDER	28
4.3.1	Schematic diagram of 8-bit CARRY SELECT ADDER	30
5.1.1	Simulated I/Ps o/p waveform of FA in STA CMOS	31
5.1.2	Simulated I/Ps o/p waveform of FA in STA PTL	32
5.1.3	Simulated I/Ps o/p waveform of FA in STA TGA	32
5.2.1	Simulated I/Ps o/p waveform of 4 bit FA	34
5.2.2	Simulated I/Ps o/p waveform of 8 bit FA	35

LIST OF TABLES

TABLE NO.	NAME OF THE TABLE	PAGE NO.
1	Truth table for 1-bit full Adder	14
2	Complete observation results of different topologies of CMOS full Adder for different supply voltage	33
3	Comparison between 4-bit and 8-bit	35

LIST OF ABBREVIATIONS

CMOS	Complementary Metal Oxide Semiconductor
PMOS	P-type Metal Oxide Semiconductor
NMOS	N-type Metal Oxide Semiconductor
FA	Full Adder
STA	Static
I/N	Input
O/P	Output
INV	Inverter
PTL	Pass Transistor Logic
TG	Transmission gate
PDP	Power Delay Product
S	Sum
CO	Carry Out
CPL	Complementary pass transistor logic

CHAPTER 1
INTRODUCTION

1.1 Introduction

The design of a Carry Select Adder is such that it operates faster than most conventional adders. The power consumed is such an adder is also moderate and a simple gate level modification is required of a regular CSA to reduce the power. Carry Select Adders are used for high speed application by reducing propagation delay. Though it requires more area than most adders but the design can be implemented in such a way that it can overcome the aforementioned difficulties in the most suitable manner. Building low power VLSI system has emerged as significant performance goal because of the fast technology in mobile communication and computation. The advances in battery technology have not taken place as fast as the advances in electronic devices. So the designers are faced with more constraints; high-speed, high throughput and at the same time consuming as minimal power as possible. They are likely to perpetuate the ability to further reduce the cost per function and improve the performance of integrated circuits. The basic operation Carry Select Adder (CSA) is parallel computation. CSA generates many carriers and partial sum [3]. The final sum and carry are selected by multiplexers. Multiple pairs of Ripple Cary Adders (RCA) are used in CSA structure. Hence, the CSA is not area efficient. The main goal of this Binary to Excess-1 converter (BEC) logic is to use lesser number of logic gate than the n-bit Full Adder. The modified CSA architecture is lower area and power consumption [10-12]. In our project, a parallel study on different types of carry select adder in 4-bit and 8-bit has been presented. They have been compared against various parameters like power, area and speed. Our survey includes: linear CSA, two stage CSA, three stage CSA, CSA with sharing and SQRT CSA. In this paper, we implement the different types of carry select adders and study the performance analysis in terms of power, area and delay using gate level (Xilinx) and circuits level (Tanner Spice) simulation tools. The rest of the paper is organized as follows: Section II gives a brief description of different types of CSA. Section III deals with the simulation results. Section IV gives the result analysis. Then comes the future work, followed by acknowledgement and conclusion.

Smaller feature dimension of transistors and higher level of integration have produced faster speed, although, with increased power dissipation and power density. High power dissipation raises temperature, degrades system reliability, and introduces high cost for heat sinks. Numerous power management techniques targeting different components of power have been proposed in the past few years. Examples of such techniques include clock- gating [1], gated-ground [2], supply voltage scaling [3], and logic and architectural level techniques [4] The relationship between dynamic/leakage power consumptions and the supply voltage (VDD) can be summarized as [3]: 2 Dyn DD ∝ PV , S_leakDDDD ~ ∝ PVV PS_leak and PG_leak denote dynamic power, sub-threshold leakage power and gate leakage power, respectively. Therefore, both dynamic and leakage power can be drastically reduced by scaling down the supply voltage [3]. 34 and DD G_leak∝ P V e . Here PDyn, The required VDD for logic operations is mainly determined by the required operation frequency and the critical path delay. Usually a margin is reserved to take into account any uncertainties in circuit/device parameters and environmental factors. However, such a worst-case-based VDD selection overestimates the actual required VDD: the combination of worst-case conditions is rare. Carry-Select Adder (CSA) provides a well-balanced choice between the slow speed of RCA andthe large area occupied by Carry Look-Ahead Adder (CLA). In this paper, we propose a novel low-power Carry-Select Adder (CSA) structure: Cascaded CSA ($C2$ SA). By distinguishing between the short- and longlatency operations, $C2$ SA works with variable latencies (1 or 2 cycles). Our design allows more aggressive VDD scaling (under the same timing constraint), or extra timing margin to tolerate the process parameter and environmental variations (under the same energy budget), while closely maintaining the same Average Latency Per Operation (ALPO) compared to standard CSA. Our experiments on a prototype 64-bit C2 SA show 40.7% and 44.4% total power savings in 180nm and 70nm technologies, respectively, under scaled VDD"s. No ALPO-based performance loss is introduced and only around 3.97% area overhead isincurred with respect to the standard CSA.

1.2 LITERATURE SURVEY

As we know adders are of fundamental importance in a wide variety of digital systems, several types of fast adders exist but adding fast using low area and power is still challenging. In digital adders, the speed of addition is limited by the time required to propagate a carry through adder. So the CSLA is used in many computational systems to alleviate the problem of carry propagation delay. So described the extremely fast digital adder with sum selection and multiple-radix carry. He compared the amount of hardware and the logical delay for a 100-bit ripple-carry adder and a carry-select adder. The problem of carry-propagation delay was overcome by independently generating multiple-radix carries and using these carries to select between simultaneously generated sums. In this adder system, the addend and augends were divided into subaddend and subagent sections that were added twice to produce two sub sums. One addition was done with a carry digit forced into each section, and the other addition combined the operands without the forced carry digit. The selection of the correct sub sum from each of the adder sections depended upon whether or not there actually was a carry into that adder section.

1.3 OBJECTIVES

- To design full adder (FA) using different logic styles.
- To design of multi bit full adder using different logic styles
- To design of carry select adder in 4-bit

1.4 THESIS OUTLINE

The Thesis is organized as follows:

Chapter 2: In this chapter 3 different full adder architecture designed and discussed.

Chapter 3: Two different multi bit FA using different logic style have been designed and discussed in this chapter.

Chapter 4: 4-bit carry select adder designed and which has been discussed in this chapter.

Chapter 5: In this chapter simulation and performance analysis of 3 different full adder 2 different multi bit FA, 4-bit Carry select adder using Tanner13.1 EDA Tool are discussed. Performance analysis is done with respect to average power consumed, propagation delay, power delay product (PDP) and transistors count are discussed.

Chapter 6: Scope of future work and conclusions are discussed in this chapter.

TOOL USED: T-SPICE

Tanner EDA tool is for analog and digital signals integrated circuits and its design offers efficient path from design captured through verification. In this we can use different applications including power management, Display image sensors, and Bio-medical, Automotive and consumer electronics. In this tanner we have several specialty tools to enhance our productivity depending upon our requirement .Different specialty tools are layer fill, Pad IO Cross-Reference Extractor (Pad Map), Wafer tools and Node Highlighting. In tanner micro electrical mechanical systems (MEMS) designs have been tapped out. Mechanical Design Tools can perform few No Boolean Operations, have no Design Rule Checking, limited scripting capability, and limited hierarchical structure so operations are slow in performing and simple layer render makes it difficult to view overlapping geometries. Compared to mechanical design tools tanner can perform enhanced Boolean operations, perform design rule checking (DRC), in this hierarchical structure makes operations fast and is easy to review overlapping Geometries

1. DESIGN ASPECTS

In this paper I am using the tanner tool to implement the convectional full adder having 28 transistors in number [4]. The process of making convectional full adder consists of mainly three steps: They are:

1. S-Edit(schematic)
2. L-Edit(Layout)
3. LVS(Layout Vs Schematic)

CHAPTER 2
DESIGN OF FULL ADDER USING DIFFERENT LOGIC

2.1 FULL ADDER

Adders are essential building blocks for multipliers. Multipliers are the elementary component and multiplication is basic operation in many digital signal processors, general purpose processors and digital filters [7]. The design of the full adder is based on the design of the XOR gate. The two O/Ps Sum(S) and Carry out(CO) can be generated based on the Boolean equations (3.1) and (3.2) of FA as follows:

$$S = a \oplus b \oplus ci \tag{2.1}$$

$$CO = b.ci + ci.a + a.b = ci.(a \oplus b) + a.b \tag{2.2}$$

Where a, b are data I/P and ci is the previous I/P carry.

Table 1: Truth Table for 1-bit Full Adder

I/Ps			O/Ps	
A	B	Ci	S	CO
0	0	0	0	0
0	0	1	1	0
0	1	0	1	0
0	1	1	0	1
1	0	0	1	0
1	0	1	0	1
1	1	0	0	1
1	1	1	1	1

The design of a different FA circuit is explained in the following section.

2.2 STA CMOS FA

Different logic styles can be investigated from different points of view. Evidently, they tend to favor one performance aspect at the expense of others. In other words, it is different design constraints imposed by the application that each logic style has its place in the cell library development. Even a selected style appropriate for a specific function may not be suitable for another one. For example, static approach presents robustness against noise effects, so automatically provides a consistent operation. The issue of ease of design is not always attained easily. The CMOS design style is not area efficient for complex gates with large fan-ins. Thus, care must be taken when a static logic style is selected to realize a logic function [13].

The CMOS structure combines PMOS pull-up and NMOS pull-down networks to produce considered O/Ps. In this style all transistors (either PMOS or NMOS) are arranged in completely separate branches, each may consist of several sub-branches. Mutually exclusiveness of pull-up and pull-down networks is of a great concern. Fig.2.2.2 shows the STA CMOS FA [13].

For STA CMOS FA 28 transistors are required.

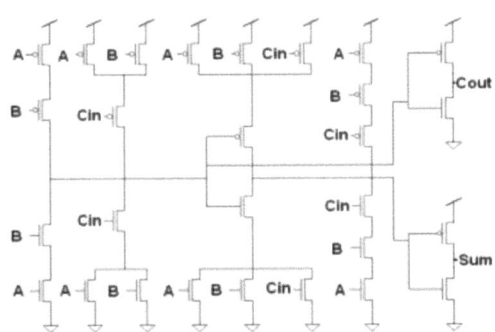

Fig. 2.2.1: Circuit diagram of CMOS FA

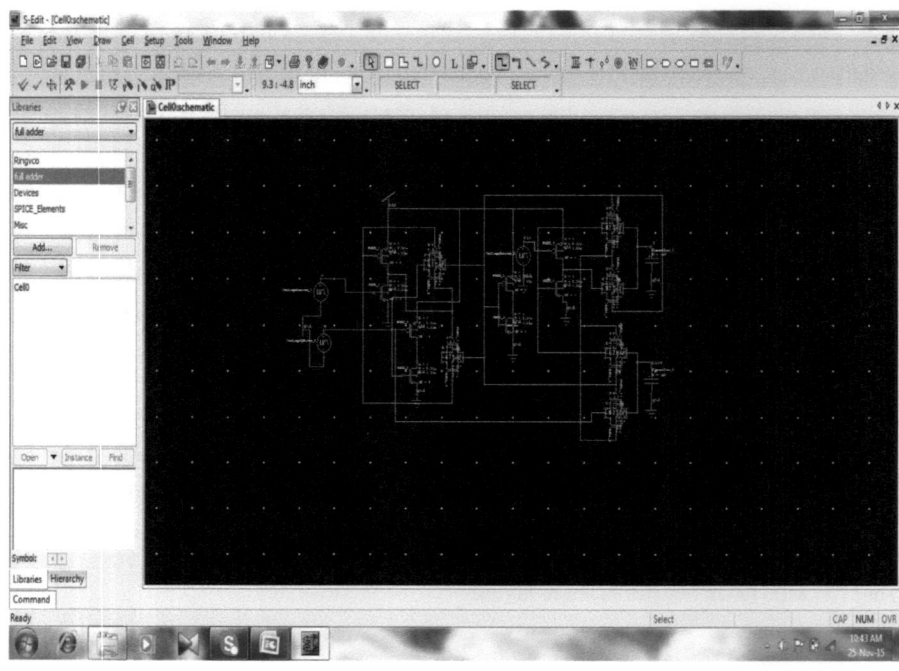

Fig. 2.2.2: Schematic diagram of standard CMOS FA

2.3 PTL FA

PTL FA is shown in Fig. 2.3.1. The major advantage of using the CMOS based PTL circuits is that the number of transistors can be reduced compared to those based on conventional CMOS. Say, only two transistors are needed for both the OR and AND gates, whereas a total of six transistors are used in the corresponding standard CMOS circuit. For XOR gate using PTL logic only 2 transistors are needed. The schematic diagram of PTL FA is shown in fig 2.3.2.

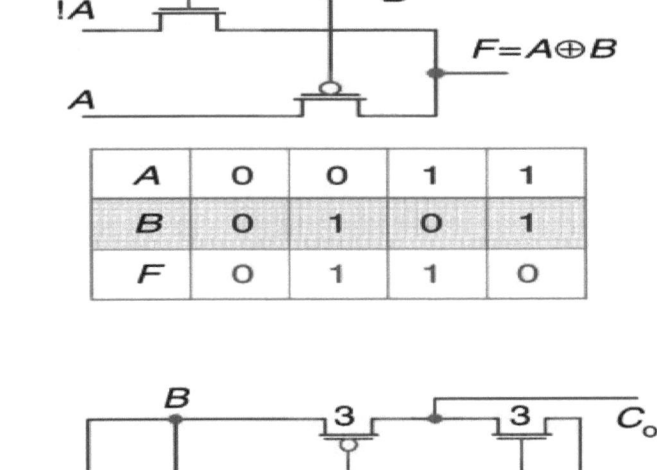

Fig. 2.3.1: Circuit diagram of PTL FA

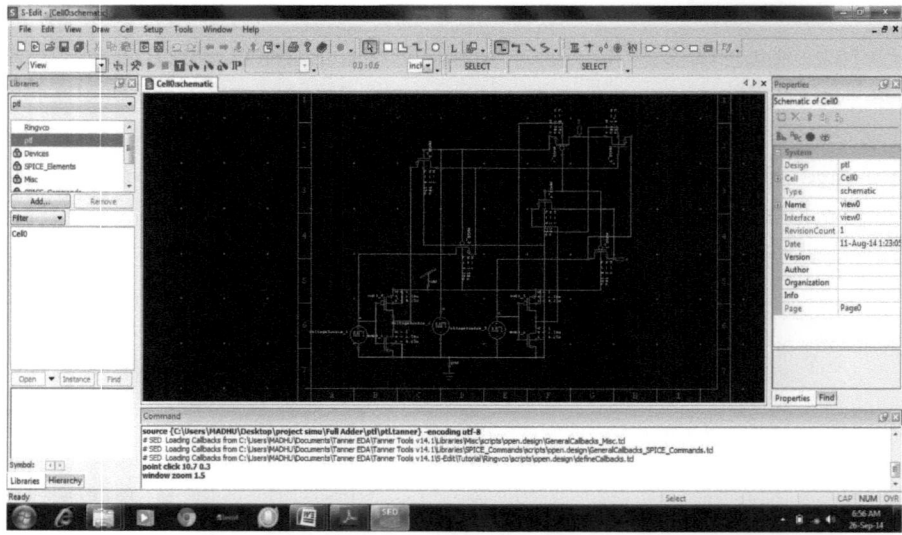

Fig. 2.3.2: Schematic diagram of PTL FA

2.4 TGA FA

The Conventional CMOS full adder is based on Transmission gates called the Transmission gates full adder cell (TGA) and it is one of the standard implementations of the 1-bit full-adder cells that is shown in 2.4.1 and has 20 transistors. Transmission gate approach is another widely used CMOS design style to implement digital function. Transmission gate based implementation is similar to pass transistor with the difference that transmission gate logic uses NMOS and PMOS transistors where as pass transistor logic uses only one type of transistor i.e. either NMOS or PMOS. The schematic view of TGA FA is shown in fig 2.4.2.

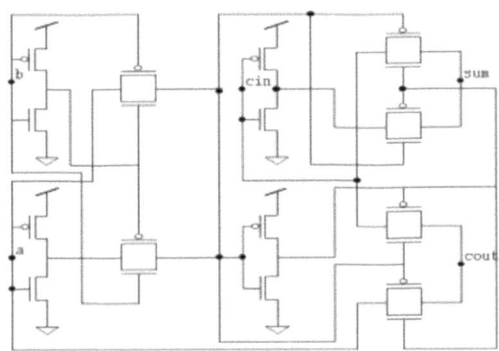

Fig. 2.4.1: Circuit diagram of TGA FA

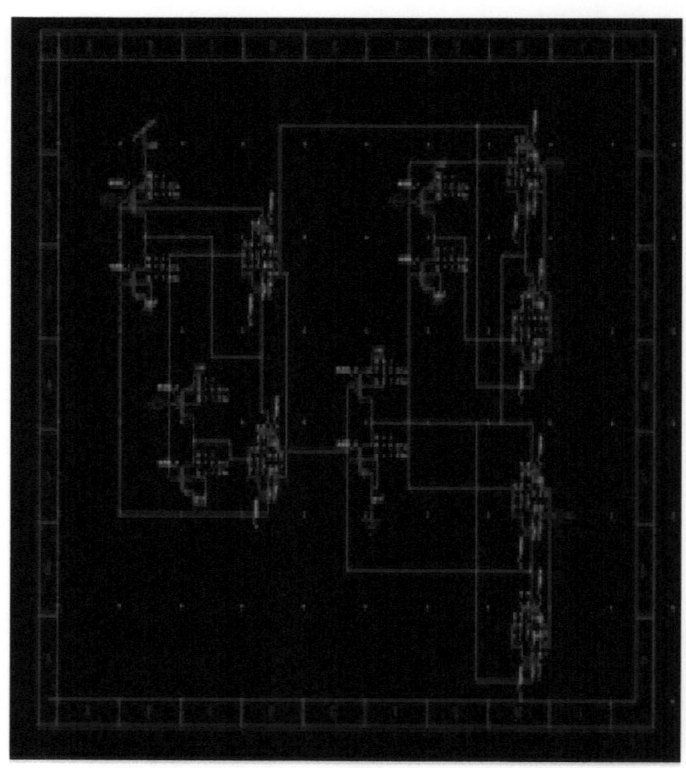

Fig. 2.4.2: Schematic diagram of TGA FA

CHAPTER 3

DESIGN OF MULTI- BIT FULL ADDER USING DIFFERENT LOGIC

3.1   4-BIT FULL ADDER

The 4-bit adder block used in CSA is ripple carry adder. In ripple carry adder each carry bit from a full adder "ripples" to the next full adder [3]. The simple implementation of 4-bit ripple carry adder is shown below in Figure 2. C0 is the input carry, x3 x2 x1 x0 and y3 y2 y1 y0 represents two 4-bit input binary numbers.C4 is the output carry and s3 s2 s1 s0 is the sum output

The ripple carry adder is designed using a full adder cell with 18-transisitors based on transmission gate logic [4]. The full adder is constructed using an XOR gate and two 2:1 multiplexers as shown in Figure 3.1.1.The SUM (A xor B xor Cin) is formed by a multiplexer controlled by A xor B (and complement). Examining the adder truth table reveals that when A xor B is true, COUT=C and SUM=complement of C. When A xor B is false, COUT=A (or B) and SUM=C. The ripple carry adder is designed using a full adder cell with 18-transisitors based on transmission gate logic [4]. The full adder is constructed using an XOR gate and two 2:1 multiplexers as shown in Figure 3.1.2.The SUM (A xor B xor Cin) is formed by a multiplexer controlled by A xor B (and complement). Examining the adder truth table reveals that when A xor B is true, COUT=C and SUM=complement of C. When A xor B is false, COUT=A (or B) and SUM=C.

It is possible to create a logical circuit using multiple full adders to add N-bit numbers. Each full adder inputs a Cin, which is the Cout of the previous adder. This kind of adder is called a ripple-carry adder, since each carry bit "ripples" to the next full adder. Note that the first (and only the first) full adder may be replaced by a half adder (under the assumption that Cin = 0).

The layout of a ripple-carry adder is simple, which allows fast design time; however, the ripple-carry adder is relatively slow, since each full adder must wait for the carry bit to be calculated from the previous full adder. The gate delay can easily be calculated by inspection of the full adder circuit. Each full adder requires three levels of logic. In a 32-bit ripple-carry adder, there are 32 full adders, so the critical path (worst case) delay is 3 (from input to carry in first adder) + 31 × 2 (for carry propagation in later adders) = 65 gate delays. The delay from bit position 0 to the carry-out is a little different: The carry-in must travel through n carry-generator blocks to have an effect on the carry-out A design with alternating carry polarities and optimized AND-OR-Invert gates can be about twice as fast.

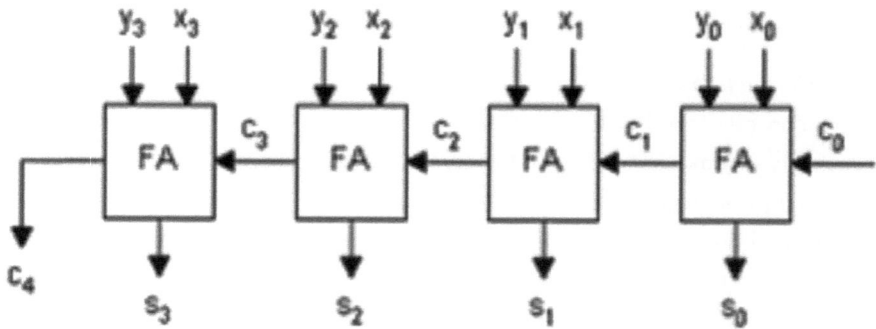

Fig. 3.1.1: BLOCK DIAGRAM OF 4-BIT FULL ADDER

Fig. 3.1.2: SCHEMATIC DIAGRAM OF 4-BIT full ADDER

3.2 8-BIT FULL ADDER

The 8-bit full adder [2] is implemented as shown in Figure 3.2.1.The adder is split into two 4-bit groups. The lower order bits a3 a2 a1 a0 and b3 b2 b1 b0 are fed into the 4-bit adder L to produce the sum bits s3 s2 s1 s0 and a carry-out bit c4. The higher order bits a7 a6 a5 a4 and b7 b6 b5 b4 are used as inputs to two 4-bit adders. Adder U0 calculates the sum with a carry-in of c=0, while adder U1 calculates with a carry-in of c =1. Both sets of results are used as inputs to an array of 2:1 multiplexers. The carry bit from the c4 of adder L is used as the MUX select signal. If c4 = 0, then the results of U0 are sent to the output, while a value of c4 =1 selects the results of U1 for s7 s6 s5 s4. The carry-out bit c8 is also selected by the MUX array.

Electronic devices such as mobile phones, cameras are used commonly these days. Its battery life span is of great concern. When mobile phone is operated in standby mode, certain programs of mobile phone or camera are turned off during active or talk mode but this doesn't stop the battery from getting depleted. This is because circuits which are de-activated by turning off certain programs still have leakage currents flowing through them. Even though the magnitude of leakage current is lesser than the normal operating current but leakage current erodes battery life over relatively long standby time whereas the normal operating current erodes battery life over relatively short talk time. Thus this is why low power circuits for mobile applications are of great interest. Implementation of adder cells to reduce the power consumption and to increase the speed has proved to be a worthy solution towards power reduction. Moreover, realization of adders with different approaches using CMOS technology widens the area of power reduction, performance of the adder cells can be evaluated by measuring the factors such as leakage power, active power in context to voltage and transistor scaling. Reducing the transistor's gate length when no voltage is applied at gate results in more leakage current between source and drain of the transistor which eventually results in the more power consumption

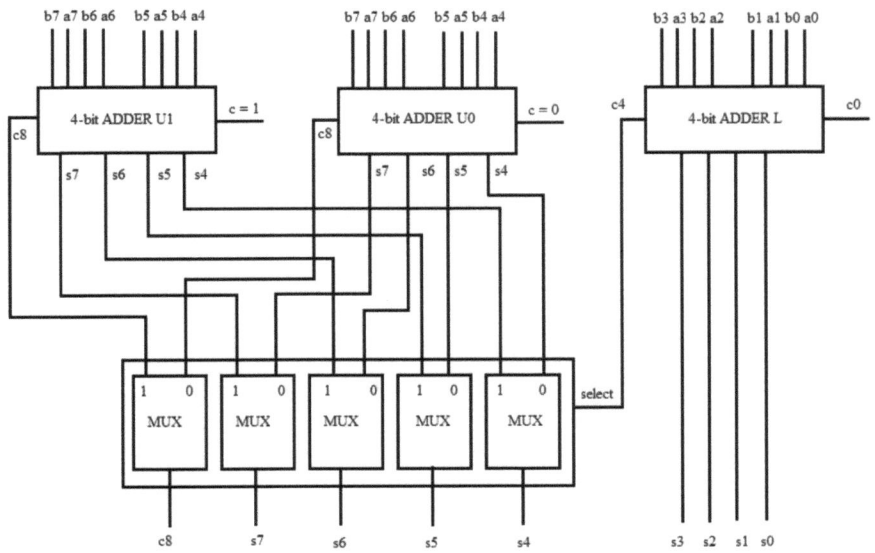

Fig. 3.2.1: Circuit diagram of 8-bit FULL ADDER

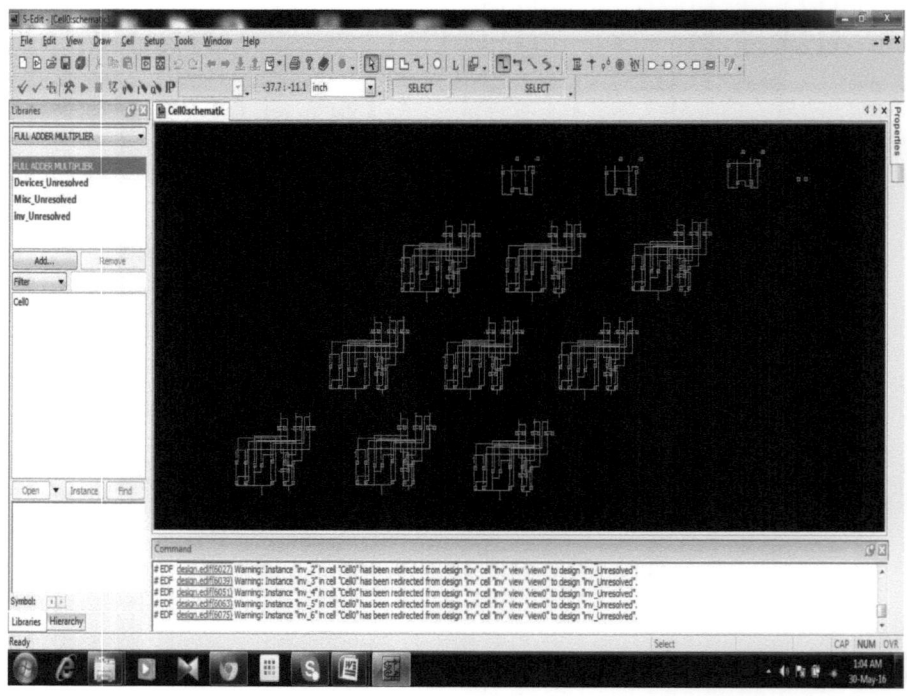

Fig. 3.2.2: Schematic diagram of 8-bit FULL ADDER

CHAPTER 4

DESIGN OF CARRY SELECT ADDER IN 4-BIT

4.1 CARRY SELECT ADDER

In electronics, a **carry-select adder** is a particular way to implement an adder, which is a logic element that computes the $(n+1)$-bit sum of two n-bit numbers. The carry-select adder is simple but fast, having a gate level depth of $O(\sqrt{n})$.

The carry-select adder generally consists of two ripple carry adders and a multiplexer. Adding two n-bit numbers with a carry-select adder is done with two adders (therefore two ripple carry adders) in order to perform the calculation twice, one time with the assumption of the carry being zero and the other assuming one. After the two results are calculated, the correct sum, as well as the correct carry, is then selected with the multiplexer once the correct carry is known.

The number of bits in each carry select block can be uniform, or variable. In the uniform case, the optimal delay occurs for a block size of $\lfloor \sqrt{n} \rfloor$. When variable, the block size should have a delay, from addition inputs A and B to the carry out, equal to that of the multiplexer chain leading into it, so that the carry out is calculated just in time. The $O(\sqrt{n})$ delay is derived from uniform sizing, where the ideal number of full-adder elements per block is equal to the square root of the number of bits being added, since that will yield an equal number of MUX delays.

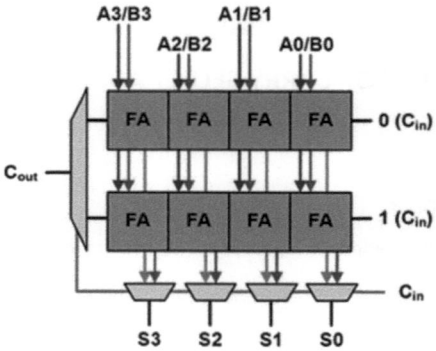

Fig. 4.1.1: Circuit diagram of 4-bit carry select adder

Above is the basic building block of a carry-select adder, where the block size is 4. Two 4-bit ripple carry adders are multiplexed together, where the resulting carry and sum bits are selected by the carry-in. Since one ripple carry adder assumes a carry-in of 0, and the other assumes a carry-in of 1, selecting which adder had the correct assumption via the actual carry-in yields the desired result.

4.2 CONDITIONAL SUM ADDER

A **conditional sum adder** is a recursive structure based on the carry-select adder. In the conditional sum adder, the MUX level chooses between two $n/2$-bit inputs that are themselves built as conditional-sum adder. The bottom level of the tree consists of pairs of 2-bit adders (1 half adder and 3 full adders) plus 2 single-bit multiplexers. The conditional sum adder suffers from a very large fan-out of the intermediate carry outputs. The fan out can be as high as $n/2$ on the last level, The carry-select adder design can be complemented with a carry-look ahead adder structure to generate the MUX inputs, thus gaining even greater performance as a parallel prefix adder while potentially reducing area.

4.3 CIRCUIT DESCRIPTION

An 4-bit **carry-select adder**, built as a cascade from a 1-bit full-adder, a 3-bit carry-select block, and a 4-bit carry-select adder. Click the input switches or type the 'a', 'b', 'c' bind keys to control the first-stage adder.

The problem of the ripple-carry adder is that each adder has to wait for the arrival of its carry-input signal before the actual addition can start. The basic idea of the carry-select adder is to use blocks of two ripple-carry adders, one of which is fed with a constant 0 carry-in while the other is fed with a constant 1 carry-in. Therefore, both blocks can calculate in parallel. When the actual carry-in signal for the block arrives, multiplexers are used to select the correct one of both pre calculated partial sums. Also, the resulting carry-out is selected and propagated to the next carry-select block.

In total, the carry propagation time through an n-bit adder block is reduced from $O(n)$ to the number of stages times the delay of the multiplexers. Naturally, using n blocks of 1-bit carry-select adders would incur a complexity of n multiplexers, again resulting in $O(n)$ delay. Therefore, a partition with (slowly) increasing block-size is chosen. In the example, the first (least-significant) block consists of a simple full adder, followed by a 3-bit carry-select block, and finally a 4-bit carry-select block. A common choice for a 16-bit carry-select adder is to use a 6-4-3-2-1 bit partitioning. While the delay of the standard ripple-carry adder with n-bits is $O(n)$, the delay through the carry-select adder behaves as $O(sqrt(n))$ at a hardware cost of $O(3*n)$.

To demonstrate this behavior, a very large gate-delay is used for the gates inside the 1-bit adders – resulting in an addition time of about 0.6 seconds per adder. The total carry-propagation time is therefore 0.6 seconds for the first adder, and another 1.2 seconds through both carry-select blocks, for a total of 1.8 seconds from the (A0,B0,Cin) inputs to the (Cout) output. The longest delay path is this circuit is through the four-bit ripple-carry block from (A4,B4) to (Cout), for a total delay of 2.4 seconds. Even in this (small) example, the carry-select adder is much faster than the ripple-carry adder at 4.8 seconds.

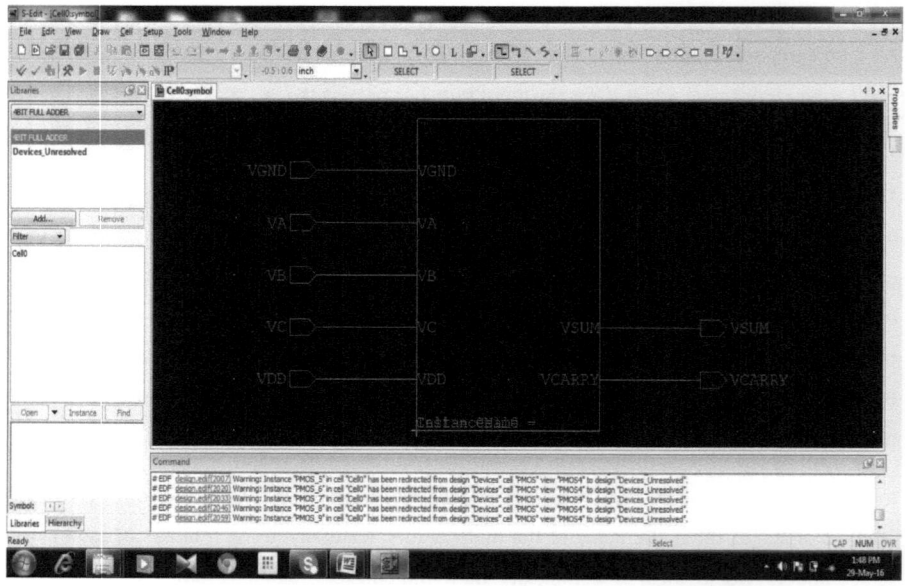

Fig. 4.3.1: Schematic diagram of 4-BIT carry select adder

CHAPTER 5
SIMULATION RESULTS

5.1 SIMULATION RESULTS FOR FA

The I/P and O/P waveform of FA is taken from Tanner simulation tool, which is shown in Fig.5.1.1 to 5.1.3. The 3 FA circuits like Std. CMOS, PTL, TGA, FAs are compared in terms of propagation delay, Average power consumption, PDP and transistor complexity. The details results are obtained by employing Tanner13.1 EDA simulation tool which is shown in Table 2.

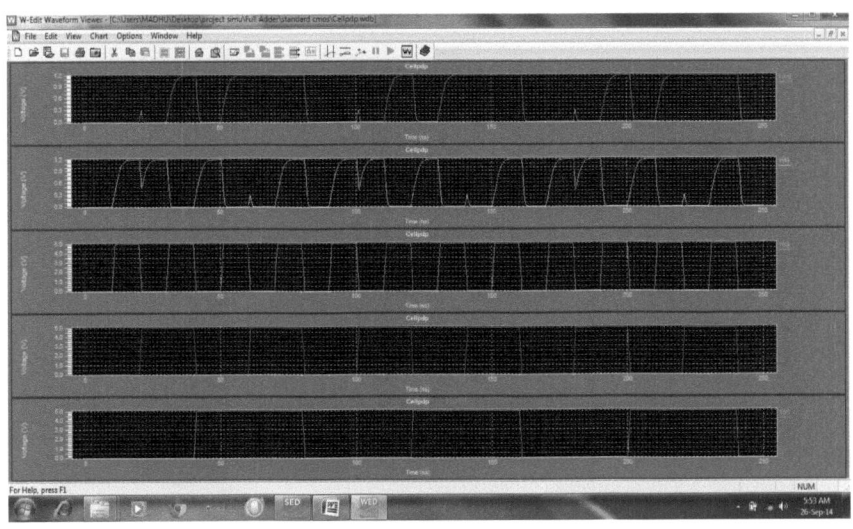

Fig. 5.1.1: SIMULATED I/Ps O/P WAVEFORM of FA IN STA CMOS

Fig. 5.1.2: SIMULATED I/Ps O/P WAVEFORM of FA IN STA PTL

Fig. 5.1.3: SIMULATED I/Ps O/P WAVEFORM of FA IN STA TGA

Table 2: Complete Observation results of different topologies of CMOS full adder for different Supply Voltages

Sl. No.	Different Topology	Variation of Supply Voltage(V)	Propagation Delay for sum (ps)	Propagation Delay for carry (ps)	Avg. Power Consumed (µW)	PDP for Sum(fJ)	PDP for Carry (fJ)	Transistor Count
1	STA. CMOS Full Adder	1.2	1286.60	1152.50	23.86	30.67	27.50	28(3.5)
		3.3	667.98	659.44	182.39	121.83	120.28	
		5.0	598.80	596.60	422.06	248.93	251.80	
2	PTL Full Adder	1.2	23.67	30.21	0.17	0.004	0.005	10(1.25)
		3.3	18.91	17.19	3.40	0.064	0.058	
		5.0	17.15	13.69	9.50	0.163	0.13	
3	TGA Full Adder	1.2	**9.34**	9.33	0.24	**0.002**	**0.002**	20(2.5)
		3.3	4.05	4.04	5.66	0.023	0.023	
		5.0	3.45	3.41	17.53	0.06	0.059	

The bold data are showing the minimum value of the different performance parameters. From Table 2 it is noted that summery of all results are given.

From Table 2, it is observed that the smallest propagation delay is found in FA for medium and high supply voltage and in TG FA for lower supply voltage. Maximum propagation delay is found in STA CMOS FA for any supply voltage. It is also noted that the smallest & highest power consumed in STA CMOS for any supply voltage. It is also noted that the smallest power delay product is obtained in for lower and medium supply voltage and in TG FA for lower supply. Maximum PDP is found in STA CMOS FA for all supply voltages.

5.2 SIMULATION RESULTS FOR MULTI BIT FULL ADDER

The I/P and O/P waveform of multi bit FA is taken from Tanner simulation tool, which is shown in Fig.5.2.1 and 5.2.2. The 4-bit FA and 8 bit FA circuits are compared in terms of propagation delay, Average power consumption, PDP and transistor complexity. The details results are obtained by employing Tanner13.1 EDA simulation tool which is shown in Table 3.

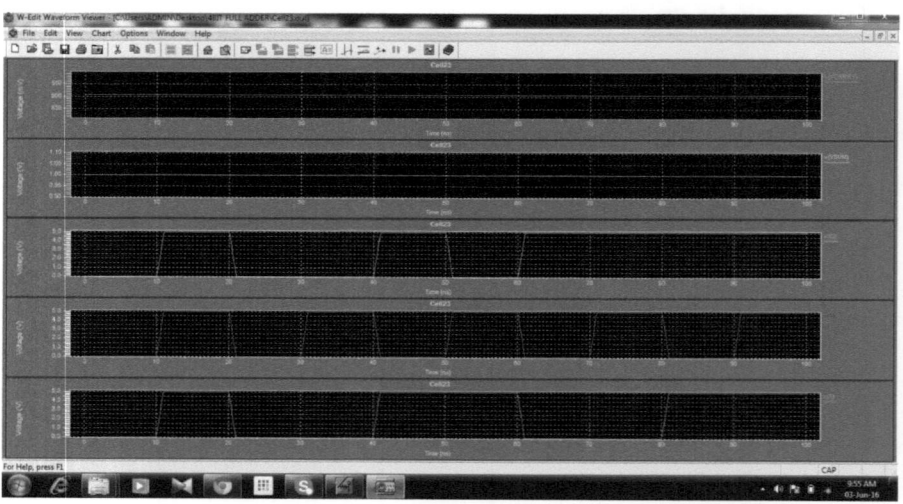

Fig. 5.2.1: SIMULATED I/Ps O/P WAVEFORM of 4-BIT FA

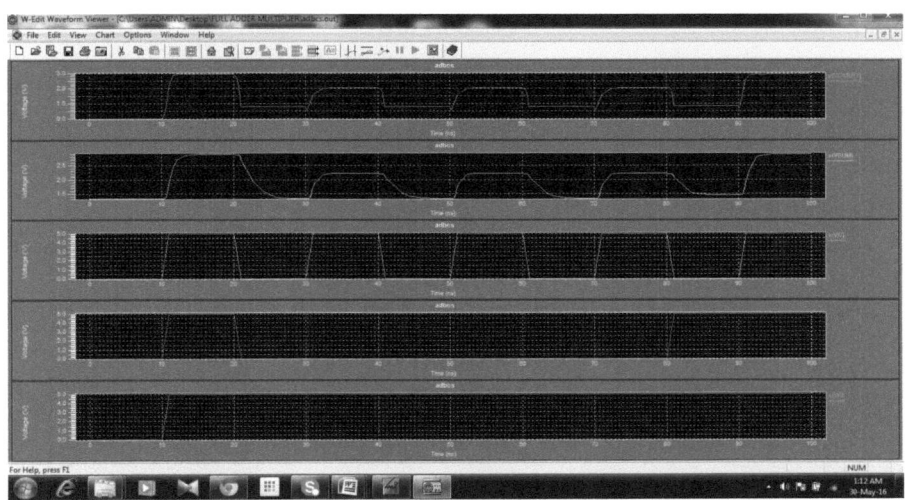

Fig. 5.2.2: SIMULATED I/Ps O/P WAVEFORM of 8-BIT FA

Table 3: Comparisons between 4 bit and 8 bit

Adder	No. Of Transistors	Propagation Delay (ns)	Power (pW)	PDP (e-21J)
4 bit	72	47.04 nsec	1.32×10^{-2} watts	620.928 pico jule
8 bit	144	10.606 nsec	2.54×10^{-2} watts	269.3924 pico jule

CHAPTER 6
FUTURE WORK AND CONCLUSIONS

6.1 FUTURE WORK

Tanner tool. Further I shall design 32-bits and 64-bits Carry -Select Adder (CSA) using obtained results and Tanner 13.1 EDA

6.2 CONCLUSIONS

In this project I have designed Full Adder by using 3 different full adder topologies like 28 transistors, 20 transistors and 10 transistors Full adder circuits are implemented using Tanner 13. 1 EDA simulation tool. They are compared in terms of Delay, power consumptions, area(no. of transistors) and Power Delay Product (PDP). While designing full adder circuits using Tanner 13.1 EDA simulation tool smallest propagation delay is obtained in 28Transistor full adder ,smallest power is consumed in PTL full adder ,smallest area in 10Transistor full adder and smallest power delay product is obtained in PTL full adder. I have also designed 4-bit and 8 –bit full adder by using cmos full adder topologies like 72 transistors,144 transistors 4-bit and 8-bit full adder circuits are implanted using tanner 13.1 EDA simulation tool. They are compared in terms of Delay, power consumptions, area (no. of transistors) and Power Delay Product (PDP). While designing full adder circuits using Tanner 13.1 EDA simulation tool smallest propagation delay is obtained in 72 transistors 4-bit full adder, smallest power is consumed in 4-bit full adder, smallest area in 72Transistor 4-bit full adder and smallest power delay product is obtained in 4-bit full adder

REFERENCES

[1] H. Li, S. Bhunia, Y. Chen, T. N. Vijaykumar, and K. Roy. Deterministic clock gating formicroprocessor power reduction. In Proceedings of the 9th Int"l Symp. on High Performance Computer Arch., pp. 113-124, Feb. 2003.

[2] Agarwal, H. Li, and K. Roy, A Single-Vt Low-Leakage Gated-Ground Cache for Deep Submicron, In IEEE Journal of Solid-State Circuits, Vol.38-2, pp. 319-328, Feb. 2003.

[3] H. Li, C.-Y. Cher, T. N. Vijaykumar, and K. Roy, VSV: L2- Miss-Driven Variable Suppy Voltage Scaling for Low Power, In Proceeds of the 36th IEEE/ACM Int"l Symp. on Microarchitecture, pp. 19-28, Dec. 2003.

[4] D. Drnst, et. al. Razor: A Low-Power Pipeline Based on Circuit-Level Timing Speculation. In Proceedings of the 36th IEEE/ACM Int"l Symp. on Microarchitecture, pp. 7-18,Dec. 2003.[5]H.

[5] Suzuki, W. Jeong, K. Roy, Low Power Adder with Adaptive Supply Voltage, *In Proceedings of the 21st Int'l Conf. on Computer Design*, pp. 103-106, Oct. 2003.

[6] J. M. Rabaey, Digital Integrated Circuits: a design perspective, *Prentice Hall*, 1996.

[7] Y. Kim and L. S.-Kim, A Low Power Carry Select Adder with Reduced Area, *In Proceedings of 2001 IEEE Int'l Symp. on Circuits and Systems*, Vol. 4, pp. 218-221, May 2001.

[8] Behnam Amelifard, Farzan Fallah, Massoud Pedram, "Closing the gap between carry select Adder and ripple carry adder: A new class of low-power high-performance adders".

[9] I-Chyn Wey, Cheng-Chen Ho, Yi-Sheng Lin and Chien-Chang Peng, "An area-efficient carry select adder design by sharing the common boolean logic term" Proceedings of the International Multiconference of Engineers and Computer Scientists 2012 Vol. II, IMECS 2012, March 14-16 2012, Hong Kong.

[10] T. Y. Ceiang and M. J. Hsiao, "Carry-select adder using single ripple carry adder," *Electron. Lett.*, vol. 34, no. 22, pp. 2101–2103, Oct. 1998.

[11] B. Ramkumar and Harish M Kittur "Low-Power and Area-Efficient Carry Select Adder" *IEEE Trans. on very large scale integration systems* 2012.

[12] Sajesh Kumar U, Mohamed Salih K. and Sajith K "Design and Implementation of Carry Select Adder without Using Multiplexers" *IEEE Conference. on* Emerging Technology Trends in Electronics, Communication and Networking.

[13] O. J. Bedrij, "Carry-select adder", IRE transactions on Electronics Computers, vol.EC-11, pp. 340-346, June1962

[14] Kuldeep Rawat, Tarek Darwish and Magdy Bayoumi, "A low power and reduced area carry select adder", 45th Midwest Symposium on Circuits and Systems, vol.1, pp. 467-470, March 2002.

[15] Sarabdeep Singh and Dilip Kumar, "Design of area and power efficient modified carry select adder", International Journal of Computer Applications (0975 – 8887) Volume 33– No.3, November 2011.

[16] K. Rawwat, T. Darwish, and M. Bayoumi, ".A low power carry select adder with reduces area", Proc. Of MidwestSymposium on Circuits and Systems, pp. 218- 221, 2001.

[17] A. Tyagi, "A reduced area scheme for carry-select adders", IEEE Trans. on Computer, vol. 42, pp. 1163- 1170, 1993

[18] Y. Kim and L-S Kim, "64-bit carry-select adder with reduced area", Electronics Letters, vol. 37, pp. 614-615, May 2001.

[19] O. Kwon, E. Swartzlander, and K. Nowka, "A fast hybrid carry-lookahead/carry-select adder design", Proc. of the 11th Great Lakes symposium on VLSI, pp.149-152, March 2001.

[20] Wang, Y. Pai, C.Song, X., "The design of hybrid carry lookahead/ carry-select adders", Circuits and Systems II:Analog and Digital Signal Processing, IEEE Transactions on Volume 49, pp.16-24, 2002